Did you enjoy this issue of BioCoder?

Sign up and we'll deliver future issues and news about the community for FREE.

http://oreilly.com/go/biocoder-news

BioCoder

FALL 2014

Beijing · Cambridge · Farnham · Köln · Sebastopol · Tokyo

BioCoder #5

by O'Reilly Media, Inc.

Printed in the United States of America.

Published by O'Reilly Media, Inc., 1005 Gravenstein Highway North, Sebastopol, CA 95472.

O'Reilly books may be purchased for educational, business, or sales promotional use. Online editions are also available for most titles (*http://safaribooksonline.com*). For more information, contact our corporate/institutional sales department: 800-998-9938 or *corporate@oreilly.com* .

Editor: Mike Loukides

Production Editor: Nicole Shelby

Copyeditor: Sharon Wilkey

Proofreader: Sonia Saruba

Interior Designer: David Futato

Cover Designer: Randy Comer

Illustrator: Rebecca Demarest

October 2014:   First Edition

Revision History for the First Edition

2014-10-15:   First Release

See *http://oreilly.com/catalog/errata.csp?isbn=9781491913321* for release details.

978-1-491-91332-1

[LSI]

Contents

Avoiding the Tragedy of the Anticommons

Mike Loukides

A few months ago, I singled out an article in BioCoder about the appearance of Open Source Biology (*http://bit.ly/radar-osb*). In his white paper (*http://bit.ly/trojok-wp*) for the Bio-Commons (*http://www.bio-commons.org/*), Rüdiger Trojok writes about a significantly more ambitious vision for open biology: a Bio-Commons that holds biological intellectual property in trust for the good of all. He also articulates the tragedy of the anti-commons, the nightmarish opposite of the Bio-Commons in which progress is difficult or impossible because "ambiguous and competing intellectual property claims ... deter sharing and weaken investment incentives." Each individual piece of intellectual property is carefully groomed and preserved, but it's impossible to combine the elements; it's like a jigsaw puzzle, in which every piece is locked in a separate safe.

We've certainly seen the anti-commons in computing. Patent trolls are a significant disincentive to innovation (*http://bit.ly/patent-trolls*); regardless of how weak the patent claim may be, most startups just don't have the money to defend. Could biotechnology head in this direction, too? In the US, the Supreme Court has ruled that human genes cannot be patented (*http://bit.ly/scotus-genes*). But that ruling doesn't apply to genes from other organisms, and arguably doesn't apply to modifications of human genes. (I don't know the status of genetic patents in other countries.) The patentability of biological "inventions" has the potential to make it more difficult to do cutting edge research in areas like synthetic biology and pharmaceuticals (Trojok points specifically to antibiotics, where research is particularly stagnant).

The Free Software and Open Source movements have done a lot to enable innovation in computing. We have a rich "commons" of software (Linux, Apache, MySQL, Hadoop, to say nothing of the many tools from the GNU project). This software commons forms the technological basis for just about every technology company in existence today, including Facebook, Google, Apple, and even Micro-soft. Can the same ideas be equally productive for biology?

I believe so. But exactly how to apply those ideas isn't clear. As tempting as the analogy is, biology isn't computing. What does (or should) open source mean for biology? We don't yet have an answer to that question. Yes, it's reasonably easy to patent or copyright a long string of As, Ts, Cs, and Gs. And for similar reasons, we could apply any of the open source software licenses to that sequence. But is that sufficient? And what does that mean? I'd like to push on those questions a bit harder.

AN OPEN SOURCE GENOME?

In computing, the notion of "open source" has a clarity that doesn't necessarily extend to biology. We know what source code means: it's a more or less complete expression of what a computer program does. The source code may be a couple of lines long, or millions, but when you run the code, the computer does what it's told to do. We don't yet have that kind of understanding in biology, and it's possi-ble we never will. It's a truism to say that DNA is a programming language that we don't understand. While we understand (to a limited extent) how DNA enco-des proteins, we're far from understanding the complexity of that mapping. One modification to DNA may have many interacting effects, some benign, some fatal. Our notion of "effects" and "side effects" confuses the issue; side effects are just the effects we don't like. As far as the organism is concerned, though, there are only effects. And we are far from understanding all the effects of any modification on all but the simplest biological systems.

So, what does it mean to say that DNA sequences are a kind of genetic "source code" for living organisms? The process by which DNA is used to build proteins is extremely complex; the code is read in both directions, furthermore, there's a logic to gene expression that we don't completely understand. If the same genetic information is present in all cells, why are some cells muscle and other liver? Genes encode proteins, and (to use a programming analogy), they're sort of like assignment statements. But you can't build a program if you only have assignments. You need conditional logic, and other control structures. We are far from understanding DNA's control structures and how they work. So, while we can call DNA a "program," open sourcing biology is qualitatively different from open sourcing a program written in Java or C. We really don't yet understand what the biological program means. What is an open source gene? What is an open source protein? Those are important questions, and we don't yet know the answers.

A COMMON LANGUAGE

Software developers have one key advantage over biologists. Software developers speak a common language. Well, more realistically, many common languages; but the differences between Python and FORTRAN are small enough that Python programmers and FORTRAN programmers can meaningfully communicate with each other. DNA may be a programming language, but that won't help us communicate if we don't understand its syntax.

As Trojok says in the white paper, "a future bio designer should be able to code the properties of a living system... by describing the desired features in a biological programming language." That programming language could be DNA, properly understood; but a better analogy might be to see DNA as the machine language, the 1s and 0s, of biology. While the pioneers of computing dealt directly with 1s and 0s, we now describe a program's "desired features" in high level languages like Python; programming in binary only happens in a few special circumstances.

I doubt that we'll end up with a single biological language; just as in computing, we will probably end up with dozens (if not hundreds or thousands). But whether there's one or many, we need those languages to exist. And we need those languages to be part of the Commons, not proprietary creations as they were in the "dark ages" of computing. Today, there are very few programming languages that don't have an open source implementation, and it's very difficult to imagine a new programming language that doesn't start as Open Source project (Swift (*https://developer.apple.com/swift/*) being a significant exception). High-level languages for biology will be the same: to succeed, they must be part of an intellectual commons. Proprietary languages are no good for sharing ideas.

ETHICS

In the last few years, we've discovered that computing isn't as clear-cut as we thought it was. In 1990, it was relatively easy to look at a program and say that we understood what it did. Now, when almost all significant applications run on complex distributed systems, tens to thousands of computers that are "in the cloud", it's much more difficult to reason about what a program can or can't do. Look at the Shellshock bug in the Bash shell: that bug may have existed when Bash was first developed, but it would have been meaningless, unexploitable. In 1989, our computer networks were primitive. We didn't have web servers, and distributed systems were exotic, experimental beasts. It was relatively simple to understand all (or almost all) of the situations in which a program could execute.

Modern computer systems are much more like biological systems than the computers of the 80s and early 90s. Both biologists and software developers have to deal with extremely complex systems, emergent behavior, and unintended consequences. Open source hasn't been immune to the problems that arise when you place software in new contexts; and biologist have to be extremely careful about the consequences of introducing unforeseen changes into organisms, or releasing organisms into the wild.

The Bio-Commons has a Bio-Ethics subgroup (*http://www.bio-commons.org/bio/ethics/*) (currently mostly empty) for discussing ethical issues. How do we manage systems that defy determininistic understanding? What do biological systems mean, and how can we use them? What responsibilities does a researcher have for his creations?

It's interesting that the Bio-Ethics group lists "the definition of individuality" as one of its concerns. Identity and individuality are certainly an important concern in software-but those issues rarely appear in the context of open source software. You write software; you apply a license; you use software in accordance with that license. What stake does individuality have in the software you write or use? Perhaps open source software and the future bio-commons can learn from each other.

SHARING

When Richard Stallman founded the Free Software foundation, his goal was to preserve the freedom to share software. Sharing was the fundamental to the culture of computing in the 1970s, but it was threatened by the shift that brought about the startup booms of the 1980s: computing itself became a commodity, and software became monetizable. Developers stopped sharing their work (in many cases, were no longer allowed to share their work) because software was something you wrapped in a package and sold. Software faced the threat of the anticommons; the free software and open source movements are a reaction to that threat. And indeed, the open source movement has won (*http://bit.ly/os-wins*).

While "sharing knowledge" has always been a scientific ideal, many outside of the sciences would be surprised just how little knowledge is actually shared. Results are locked up in journals, which live behind carefully maintained (and extremely expensive) paywalls. Papers share results, but rarely share the actual data or the software used to analyze the data. Papers describe experiments, but rarely describe them accurately enough for their results to be duplicated reliably.

As we're engaging in research, we need to share data, we need to share code, we need to share experimental designs. But we don't yet have standard languages for sharing that information, or repositories in which to store it. Much of the data collection in the sciences is fairly haphazard. We're limited by tools and methodologies that were developed when data was hard to get and data storage was expensive. Now that you can buy terabyte disk drives for a few dollars (this morning, I see a 3 Terabyte external disk drive for US$120 retail), and fill those disk drives using automated instruments controlled by an Arduino or Raspberry Pi, we have the ability to generate and store data in bulk. We have the ability to instrument and monitor every stage of an experiment in detail; but that's not happening in biology, at least not on a regular basis.

This is an area where biologists can learn from software developers. Modern software systems throw off gigabytes of data, and we have built tools to monitor those systems, archive their data, and automate much of the analysis. There are free and commercial packages for logging and monitoring, and it continues to be a very active area of software development (*http://bit.ly/i_heart_logs*), as anyone who's attended O'Reilly's Velocity conference (*http://velocityconf.com/*) knows.

One critical goal of the Bio-Commons is to facilitate sharing. And I'm excited that they realize how little we know about sharing. We can talk about "open source" biology, but we don't really know what we mean. Are we talking about some genetic code? Are we talking about proteins? Are we talking about experimental procedures (protocols)?

In addition to the Bio-Commons, we see startups like Synbiota (*https://www.synbiota.com/*) working on cloud-based repositories for storing and sharing biological data (*http://bit.ly/lst-cloud*), much as GitHub serves as a repository for source code.

TOOLS

I've often said that the revolution in biology depends on a revolution in tooling. That revolution is also under way; I've come across many startups working on tools for biologists, ranging from the extraordinarily ambitious to the humble, and looking at customers from huge industrial laboratories to small bio-hacking spaces.

Again, it's important that the tooling biologists use be part of the biological commons. You can see it in software projects like Cytoscape (*http://www.cytoscape.org/*) and BioPython. You can also see the tooling revolution in the OpenPCR project (*http://openpcr.org/*), the low-cost homebrew PCR described in this issue of BioCoder (*http://biocoder.oreilly.com/*), and the open sourced laboratory robotics platforms from Modular Science (*https://www.modularscience.com/*) and OpenTrons (*http://bit.ly/open-trons*).

FERMENTING REVOLUTION

We're making tremendous progress in our understanding of life; we're clearly at the start of a revolution in biology. But for that revolution to get going in earnest, and to avoid settling into a dystopian anti-commons, we need to improve our ability to share. The computer revolution arguably started in the 1960s, but it really didn't get going until we understood the importance of shared code. The biological revolution will be similar, but with one big advantage: we can see what the Open Source movement has done. Many of the problems we face have already been solved, or are being solved.

We are building a biological commons. Whether that's the Bio-Commons that Rüdiger Trojok and his collaborators are building, or something that hasn't yet started to take shape, its time has come. It's the fermentation vessel chamber in which the revolution will grow.

PCR for Everyone, Everywhere

Ezequiel Alvarez-Saavedra and Sebastian Kraves

Abstract

The polymerase chain reaction (PCR)[1] is used by millions of scientists around the world. PCR enables DNA analysis that is vital for medicine, research, forensics, agriculture, and more. Despite PCR being an essential, three-decades-old technology, PCR instruments have remained relatively inaccessible. They are bulky, heavy, and usually cost several thousands of dollars. The device interface is often confusing to occasional users. We have developed a small and intuitive PCR machine (*miniPCR*) that is accessible to experimental scientists working in mainstream laboratories as well as in nontraditional settings. miniPCR units are already being used by academic biology researchers, science teachers, and independent/citizen scientists. The goal is to increase access to DNA technologies by scientists, educators, and enthusiasts around the globe.

Background

The PCR was invented by Dr. Kary Mullis in 1983. In the more than 30 years since it was first conceived, PCR has revolutionized biology and forever transformed our relationship with DNA. Dr. Mullis was awarded the Nobel Prize in Chemistry for PCR in 1993.

PCR is an *in vitro* (in a tube) process that copies (amplifies) a specific sequence of DNA in a biological sample. PCR amplifies DNA exponentially, giving scientists virtually unlimited amounts of precisely targeted genetic material in a very short time, usually two hours or less. The starting sample can be as little as a single, small DNA molecule such as a plasmid, or as complex as the genomic DNA from a whole organism. Because PCR is so powerful yet experimentally simple, it can be used in a broad range of applications, from detecting DNA specifi-

---

1 See the PCR page at the DNA Learning Center (*http://bit.ly/dna-center*) for more information.

cally (for example, diagnostic testing for genetic or infectious disease) to engineering synthetic DNA molecules (for example, cloning amplified fragments into recombinant DNA molecules). It is almost impossible to imagine the practice of modern biology without PCR.

For all its conceptual simplicity and experimental robustness, our ability to do PCR is still tied to the accessibility of PCR machines. While an eager experimental biologist armed with abundant patience and three water baths could do PCR, the process is greatly facilitated by *PCR machines*—also known as *thermal cyclers*—that automate DNA amplification. Well-funded laboratories have no shortage of thermal cycler options in the commercial PCR market; these are typically high-end users able to pay up to $10,000 for a single piece of equipment. Yet there is still a need for the broader scientific community to have unrestricted access to the convenience and power of PCR. This includes professional and independent scientists, learners, and troves of biocurious folks avid to experiment with DNA.

Making PCR Machines for Everyone

PCR machines work by heating and cooling the experimental samples. In order to amplify DNA, one must subject a biological sample to precise temperature steps, and rapidly repeat the steps over many *thermal cycles*. In the first temperature step, the two strands of the DNA double helix are physically separated by high temperature in a process called DNA *denaturation*. In the second step, the temperature is lowered very precisely so that the target DNA region can *anneal*, or pair up specifically with synthetic DNA oligonucleotide primers. Temperature is again raised for the third and last step, called *elongation* or *extension*, where a heat-stable DNA polymerase elongates the primers into a full copy of the target DNA by adding nucleotides, the building blocks of DNA.

If everything goes as expected, every time a denaturation-anneal-extension cycle is completed, each of the two strands of DNA is copied, doubling the amount of target DNA. Theoretically, 30 cycles would result in 2^{30} copies of starting material (that's about one billion molecules).

In most modern PCR machines, a thermoelectric (Peltier) element drives temperature changes in a metal block with wells that host the experimental samples. Other thermal cyclers operate by moving the reaction tubes through multiple blocks, each of which is kept at a constant temperature. So when all is said and done, thermal cyclers are just very utilitarian (if very precise) heating and cooling devices. Despite their conceptual simplicity, thermal cyclers have remained very expensive for decades, typically priced from $3,000 to $10,000. There are exceptions to this rule; for example, OpenPCR offers a kit containing about 200 parts to assemble a thermal cycler for $599.

Ready-to-use thermal cyclers are still beyond the reach of many research, teaching, and independent science labs. Insufficient thermal cycler capacity, even in well-funded labs, leads to reduced productivity and frustration. In addition to the affordability barriers, accessibility is further compromised by the design of existing PCR instruments: they are large, bulky, and hardly intuitive to use. We believe that the limited accessibility of thermal cyclers represents one of the biggest bottlenecks to a complete bio revolution.

Our goal was to design, develop, and manufacture a thermal cycler that could bring PCR to everyone. The project started as a collaboration between Ezequiel Alvarez Saavedra and a team at Templeman Automation that included Chris Templeman, Sean Jeffries, Cameron Dube, Dave Thomas, Michael White, and Randy Creasi, with contributions from Mac Cowell. About a year later, Ezequiel and Sebastian Kraves joined forces to start Amplyus, an effort to make lab science more accessible. Together, we continued development of the hardware and software until a novel PCR machine could be brought into production and put in the hands of every user. We called it *miniPCR*.

As a first step, we pondered the basic functions of a thermal cycler and considered feasible technology solutions for each step in the process. The premise was that these solutions be energy-, space-, and cost-efficient.

Figure 2-1 illustrates the basic functions of a PCR thermal cycler, and below we describe the design choices the team made for each of them.

Heating
We sought a design choice that would lower the thermal mass of the block without negatively impacting temperature stability. We found that given the proper block geometry and surface-to-volume ratios, we could utilize thin-film heaters for raising and maintaining the temperature of the samples as rapidly and precisely as with most state-of-the-art PCR devices.

Cooling
In conventional desktop thermal cyclers using Peltier technology, large heat sinks and fans are employed as a means to dissipate the heat generated in the thermoelectric cooling process. Instead, we utilized high-volume, high-capacity fans to rapidly lower block temperature through convective cooling. This design utilizes much less power and space than existing systems and can be run off batteries for use in the field.

Cycling
A microprocessor receives input from a temperature sensor and calculates the amount of heating or cooling required to reach the target temperature using a PID controller. PID controllers calculate the difference between the current and the desired temperature and attempt to minimize that difference, taking

into account the past and predicted future rates of change. This allows for fast temperature changes to be maintained stably after the target temperature is reached. PID control is used in industrial ovens, air-conditioning systems, injection molding machinery, and more. For programming flexibility and to keep the system open for future enhancements, the team built the embedded firmware on an Arduino platform.

Programming and user interface

User interfaces built into PCR machines can be cumbersome and confusing to operate, especially those at the lower end of the price spectrum. Creating, saving, storing, and then finding PCR programs can be especially time-consuming. Given our increasing reliance on computers, smartphones, and tablets, we decided to utilize these portable user interfaces and connect them to the PCR machine via USB. An intuitive software application makes it easy to program PCR protocols, and also to monitor reaction parameters during and after a run. On the PCR machine itself, a minimalistic panel of three LEDs displays the machine status.

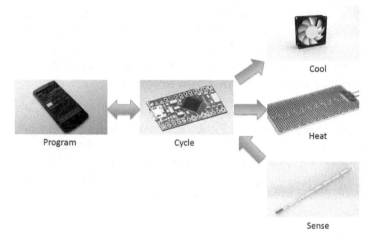

Figure 2-1. PCR engineering kept simple. (GrabCAD images used with express permission from their authors.)

The product of this design and development effort is a thermal cycler that offers the same results as other PCR instruments, and also maintains standard PCR tube formats used universally across labs. The intuitive user interface (a single hardware button, or full control via mobile or desktop app) is intuitive to novice and expert users alike. Because of its small size (the footprint of a smartphone and just one pound in weight), it changes the paradigm of where PCR belongs.

For scientists, thermal cyclers are sometimes relegated to an equipment room far away from the personal working bench; now we can have a dedicated PCR machine right where science happens. Schools can provide each group of students in science classes with their own thermal cycler without cluttering their work surfaces. It also enables teachers, scientists, and epidemiologists to move PCR machines easily from lab to lab and even take them during field trips. And it can be tucked away in a drawer when not in use, freeing up precious bench space.

miniPCR units come fully assembled and ready to use, and include Windows, Mac, and Android software. The eight-tube *mini 8* is shown in Figure 2-2; it can be purchased for $799 and has a one-year warranty and the full technical and scientific support of a dedicated team of PhD scientists.

Figure 2-2. miniPCR

miniPCR Use Cases

While the general goal of this project is to enable everyone to do cutting-edge biology, we developed miniPCR with three specific goals in mind:

1. Enable molecular biology education
2. Bring cutting-edge science to more distant places
3. Enable independent science at home

MINIPCR FOSTERING SCIENCE EDUCATION

This summer, the miniPCR team supported the Whitehead Institute's CampBio, where more than 30 middle-school "DNA scientists" from sixth through eighth grades learned how researchers work to answer some of biology's most challenging questions. These middle-school scientists used molecular detection technologies (PCR, restriction analysis, and gel electrophoresis) to help identify and control an outbreak of *E. coli*. Small groups of students worked directly with their own miniPCR machines and learned firsthand how to effectively use real molecular biology equipment. Students also had a hands-on opportunity to internalize the theory behind the PCR laboratory technique by interacting with miniPCR software during the reactions. Students were captivated by the real-time projection of the "PCR trace" on the screen: a live plot of temperature over time. They counted the exponential amplification of DNA molecules PCR-cycle by PCR-cycle, to create a 2-to-the-30th-power choir.

After this and similar teaching experiences, feedback from educators consistently reflects on *the power of streamlining and simplifying a complicated piece of equipment for student use.* Students directly comment on the staying power of active hands-on learning: "My favorite part of the program was the PCR because there was the most hands-on activity. Looking at things under a microscope is great, but it doesn't feel like we're doing anything." Figure 2-3 shows kids preparing the experimental DNA samples for analysis in the miniPCR machine. Figure 2-4 shows the user-friendly software interface that students can use to monitor the reaction and understand the underlying science.

Figure 2-3. Middle-school students learning to do experimental DNA science with miniPCR at the Whitehead Institute's CampBio program

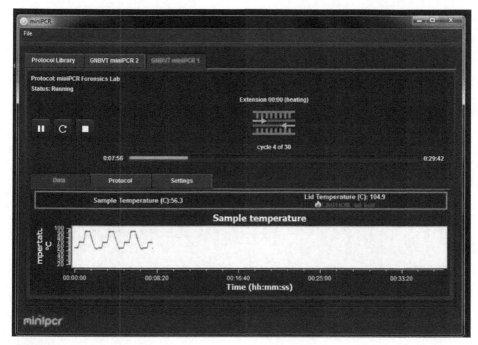

Figure 2-4. Real-time plot of the PCR process on miniPCR software

MINIPCR TAKING DNA SCIENCE GLOBAL

miniPCR's most recent global destination has been Haiti, where it joined the Haitian Bioscience Initiative (HBI) in the spring of 2014. The HBI seeks to train young Haitians in modern scientific laboratory techniques that can be helpful in environmental monitoring, specifically to ensure food and water safety, which has historically been challenging in the country. American and Haitian educators teamed with Haitian interpreters to deliver hands-on training to local students, imparting learning experiences relevant to public health in Haiti. This included microbiology testing by PCR. For the experimental techniques to be feasible, complete labs had to be assembled, in many cases with equipment flown in especially for the occasion by US-based volunteers. Bringing miniPCR to Haiti was as simple as sticking one in a carry-on backpack, and plugging it in at the other end to deliver hands-on training on molecular food-safety testing.

Students enjoyed the experimental aspects of PCR as well as the conceptual reinforcement enabled by the software, as shown on the projector screen in Figure 2-5. We have plans of fostering deeper ties with the Haitian bioscience education community.

Figure 2-5. Molecular food safety learning lab with miniPCR in Haiti

MINIPCR ENABLING INDEPENDENT SCIENCE

After obtaining PCR equipment, the next barrier that independent or under-resourced scientists face is the high cost of reagents and consumables necessary for experimental DNA science. This naturally includes plasticware, enzymes, synthetic DNA oligos, and buffers, but interestingly, there's also a perceived generalized need for special water for PCR experiments. This "PCR water" comes in the shape of *ultrapure*, *molecular grade*, and *PCR grade*, and can cost over $5,000 per liter![2] These waters are typically certified to be free of nucleases, so that DNA products do not get degraded, and to be free of nucleic acid contamination, so that the PCR does not yield false positives. In addition, purified water is *ion free* (that is, plain H_2O); this seeks to minimize interference with the activity of the DNA polymerase. Purchasing ultrapure or PCR-grade water might be justified for some experiments—for example, when attempting to detect naturally occurring nucleic acids that are ubiquitously expressed, such as those coding for 16S ribosomal RNAs. But this precaution might be excessive for a vast number of PCR applications, such as when amplifying synthetic DNA sequences from plasmids or other sources. For example, when trying to generate glow-in-the-dark plants (*http://www.glowingplant.com*) or bacteria that smell like bananas (*http://bit.ly/banana-*

---

2 $89.75/10ml = $8,975/l (*http://bit.ly/pcr-am9935*), $67/10ml = $6,700/l (*http://www.mobio.com/pcr-water/*), and $137/25ml = $5,480/l (*http://bit.ly/pcr-water*).

bac), the sequences amplified are so specific and rare in nature that there is little risk of contamination with any exogenous DNA that may be present in the water.

We wondered whether the use of regular "open source" (read: tap) water would impact PCR outcomes. Back in our academic research labs, this experiment wouldn't even have crossed our minds (at least we would not have told anyone about it). Equipped with miniPCR, we set out to conduct an assessment of the feasibility of using tap water in PCR.

We started by identifying experimental "open source" water. Living in Somerville, Massachusetts, our communities are supplied by the Massachusetts Water Resources Authority (MWRA). The MWRA obtains its water supply from the Quabbin Reservoir, the Wachusett Reservoir, and the Ware River, which have a combined capacity of approximately 477 billion gallons.[3] According to the Boston Water and Sewer Commission, "the typical customer pays just over a penny per gallon" of water supplied by the MWRA. The water for the neighboring city of Cambridge comes from a different source, the Hobbs Brook and Stony Brook reservoirs. We splurged and also tested regular bottled water purchased at a local grocery store. The positive control was double-distilled, Milli-Q filtered water, kindly donated by a local research lab. Even though Milli-Q water is not considered by some as clean as PCR-grade water, many labs use this water for their everyday PCR needs.

We used the different waters to amplify a 400-bp region of a plasmid (pMAL-c5E, New England Biolabs) through 30 cycles of amplification. In every case, test water accounted for 72 percent of the volume of the PCR mix (18 of 25 microliters). To challenge the robustness of PCR and reveal potential differences between the waters, we used stringent conditions under which DNA amplification would barely work. We used denaturing, annealing, and extension times of just 5 seconds for each cycle.

The result, likely surprising to many, is that all waters yielded PCR results! As you can see in Figure 2-6, samples with the Milli-Q water had the greatest amplification, but all other waters gave clear bands. It is likely that less restrictive PCR conditions would result in comparable amplification across samples. Another inexpensive source of potentially cleaner water is the distilled water sold in gallon jugs at grocery stores (for example, Market Basket, $0.99 per gallon).

3 See the Boston Water and Sewer Commission website (*http://bit.ly/bostonh2o*) for more.

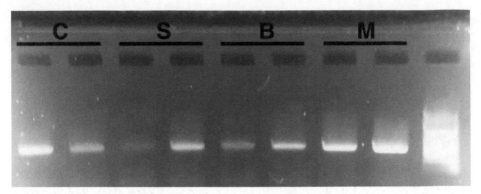

Figure 2-6. Gel electrophoresis on a 1 percent agarose gel. Samples were run in a miniPCR thermal cycler in duplicates (two different PCRs). C = water from Cambridge public water supply; S = water from Somerville public water supply; B = bottled dinking water; M = double-distilled, Milli-Q purified water. The rightmost lane is 100 bp DNA ladder (New England Biolabs). Pictures were obtained with a smartphone camera.

Several PCR applications, such as reverse-transcriptase PCR (in which the starting material is RNA), PCR used in microbiology for the specific detection of microorganisms, or "detection limit" PCR, will of course require the use of sterile and clean sources of water (RNAse and DNAse free, DNA free, autoclaved, and UV irradiated, etc.). But we can confidently recommend the use of widely available, "open source" tap water for the vast majority of nondiagnostic PCR applications. In many of the contamination-sensitive cases, the addition of *blank* tubes omitting the DNA template can provide the right level of negative control to detect potential contaminants in the water.

Conclusions and Future Directions

Our ultimate goal is to make DNA science accessible for everyone, everywhere. The motivation is personal; we are molecular biologists who had limited access to biotechnology tools while going to school and choosing a career. We are excited to be helping put PCR in the hands of more independent scientists, students, and DNA enthusiasts.

As we quickly realized when the core engineering efforts were winding down, making miniPCR available was not the end of the game but just the beginning. Our efforts to enable experimental DNA science go beyond lab technology. We collaborate and volunteer actively with science outreach groups, for example, the Bay Area Biotechnology Education Consortium (BABEC) (*http://babec.org/node/151*); MassBioEd (*http://bit.ly/massbioed*); Harvard Life Sciences Outreach (*http://*

outreach.mcb.harvard.edu); and Rockefeller University (*http://bit.ly/rock-u*). We have also started a three-way collaboration (*http://bit.ly/highschool-pcr*) with MassBioEd and New England Biolabs to bring content, technology, and reagents into the hands of teachers eager to implement biotechnology labs.

There are now over one hundred miniPCR machines in labs and homes, and we're focused on lowering cost further to make them even more accessible. We will also be taking other essential laboratory technologies and reinventing them to be smarter, more accessible, and more engaging, to transform the way we learn and teach DNA science. We want to help enable the next scientific revolution.

We are based in Cambridge, Massachusetts, and can be reached at team@minipcr.com or through our website, http://www.minipcr.com. We'd love to hear from you! Please get in touch.

Ezequiel Alvarez-Saavedra cofounded Amplyus to help make science simpler and more accessible to more people. Zeke has conducted biomedical research alongside two Nobel laureates and is the author of multiple publications in top peer-reviewed scientific journals. His work has been cited thousands of times and profiled in national and international media such as the New York Times, National Public Radio, and the BBC. He is also an inventor of gene-detection technologies. Zeke studied biology at the University of Buenos Aires, obtained his PhD in biology from MIT, and holds a BSc from Stanford University. In his spare time, Zeke digs soil in search of new species (one so far!) and is learning to play guitar alongside his son. Zeke can be reached via email at zeke@minipcr.com.

Sebastian Kraves cofounded Amplyus to help bring science to more people in more places. He previously worked on making biomedical technology accessible with the world's leading philanthropies, corporations, and multilateral organizations. As a molecular neuroscientist, Sebastian has published widely cited work on neural circuits, optogenetics, and the molecular regulation of circadian behavior. He obtained his doctorate in neurobiology from Harvard Medical School, was a postdoctoral fellow at Harvard University, and studied economics and biology at the University of Buenos Aires. Sebastian enjoys twisting into a pretzel on the yoga mat, sometimes alone and other times with his daughters. You can email Sebastian at seb@minipcr.com.

miniPCR and the miniPCR logo are trademarks of Amplyus, LLC.

Bioreactors and Food Production

Gregory Mueller

A *bioreactor* is a device or system that controls a biologically active environment. Bioreactors have been around for decades in different forms, yet with new fabrication tools and web-connected microcontrollers, makers and biohackers are starting to design modular reactors that are low cost and robust in their ability to control a biological process. There are a ton of potential applications for bioreactor technology. Here I'm going to talk about food production: specifically, home fermentation, hydroponics, and fertilizer. Then I'll share some of the reactors I've developed.

Home Fermentation

Home fermentation has gained a lot of popularity lately. Brewing at home tends to produce unique flavors, while also preserving vitamin, enzymatic, and biological activity. Plus, it's really cool! You can experiment with a wide array of things, beer probably being the most popular. Beer brewing is a two-step process of first converting a starch source into sugars (mash), and then fermenting those sugars into alcohol. Wine fermentation is fun as well, using grapes (or other fruit) as the sugar source. The primary fermentation uses yeast to convert the sugars to alcohol. The secondary malolactic fermentation uses specific strains of bacteria to convert malic acid to lactic acid for a softer taste. Micro-oxygenation is also often done to polymerize tannins, creating a smoother mouth feel.

Maybe more interesting than well-known ferments like wine or beer are probiotic sodas such as kombucha and water kefir. These are fermented by a symbiotic culture of bacteria and yeast (SCOBY). Kombucha feeds off sugar and tea, while kefir grains need only sugar water. These fermented sodas are fun because you can use all types of flavorings in a secondary fermentation. They tend to be more approachable because the end product is an effervescing healthy soda rather than an alcoholic beverage. The SCOBY, or *mother* as it is sometimes called, is also really interesting as this is absent in traditional beer or wine brewing. The SCOBY, which looks like a gelatinous disc in kombucha and like grains of quartz in water kefir, can be continually used over and over again, sometimes lasting for years. Once you begin experimenting with fermentations like this, you'll start tumbling down a bit of a rabbit hole, which is always good! For ideas on the vast world of fermented foods and beverages, check out Jen Harris's Farm to Fermentation Festival (*http://www.farmtofermentation.com*), and for starter cultures, check out Cultures for Health (*http://www.culturesforhealth.com*).

Bioreactors for fermented foods could technically be as simple as a half-gallon ball jar with an air lock, sugar, water, and kefir grains. Yet things become much more interesting when you build a more sophisticated brew setup with a fermentation controller to monitor and control temperature, pH, and other biological factors. These controllers are intriguing, not only for quality control in larger batches, but also for fine-tuning your fermentation to produce varied flavors and nutrient content!

Hydroponics

Building on this theme, hydroponics is gaining a ton of interest as well. People are developing intelligent food-production systems that are modular and being set up anywhere from backyards to rooftops. In hydroponics, the plants get all of their nutrients from the water, with the roots growing in an inert medium such as coco coir (coconut husk), perlite, or the water itself. Aquaponics incorporates the culturing of aquatic animals, predominantly fish, into the hydro system to act in a symbiotic relationship with the plants, in which the fish excrement becomes nutrients for the plants. These systems are bioreactors in their essence and are most effective when they develop a healthy microbial biofilm of nitrogen-fixing bacteria that breaks down the fish waste into nitrates/nitrites and prevents toxic by-products from accumulating.

Similar to the fermented beverage reactors, hydroponic systems can benefit greatly from a controller that allows you to govern the equilibrium of the growing system through an array of sensor and actuator feedback mechanisms. You can do cool things like choose "recipes" for what you want to grow, and the hydro system will take care of it. That's the goal, at least, and a lot of people are working on great projects like this.

Fertilizer

Fertilizer is another bioreactor application in the food world that isn't quite as visible as homebrew and hydroponics, probably because it's more difficult for the average person to participate. Bioreactors can be very effective at breaking down organic waste into nitrogen and microbially rich fertilizers. Biodigestion is an interesting four-stage fermentation, including hydrolysis, acidogenesis, acetogenesis, and methanogenesis. All are governed by different types of bacteria. Methanogens, or *archaea*, complete the process, and have recently been categorized into their own branch of the phylogenic tree of life, alongside bacteria and eukaryotes (multicellular organisms). Fascinating stuff, but that's for another article!

Most biodigesters are purely anaerobic to encourage full digestion of the feedstock, allowing optimal production of methane as a renewable energy source. The anaerobic process prevents off-gassing of nitrogen, unlike aerobic composting, which can be extremely valuable given that nitrogen is by far the most important fertilizer in modern-day agriculture. So much so that almost 1 percent of manmade energy goes to the Haber-Bosch process, which uses extreme pressure to break the triple-bonded nitrogen in the air and repurpose it into ammonia fertilizer. This is very energy intensive, and digesters could be an alternative. The downside is that the leftover nitrogen-rich digestate can be toxic as a fertilizer, filled with volatile organic acids and anaerobic organisms that aren't good for your crops. However, much like with fermented beverages and hydroponic growing systems, using a controller in your reactor to precisely modulate aeration, temperature, pH, and redox, makes it possible to create a chemically and biologically healthy fertilizer from this process.

Reactors

There are so many possibilities for bioreactors and this just begins to scratch the surface. A lot of the new Silicon Valley-style food startups are developing bioreactor and fermentation/extraction technology to make plant/microbe/algae-based protein foods and nutrient supplements. Pretty wild!

Here are a few reactors that I've developed:

Figure 3-1 shows my most recent reactor, a web-controlled device that I showed at Maker Faire '14 in San Mateo with Jen Harris from the Farm to Fermentation Festival. We developed a web interface to allow kids and anyone else to play around with the controls, including temperature, lighting, air, and dosing. This device is a fun way to make small batches of fermented beverages, running experiments with the plug-and-play sensor port, and learning about biology in general.

The reactor seen in Figure 3-2 is powered by a Linux board running a Node.js application, which controls a custom sensor/actuator array and serves a local user interface. It also connects to a remote application, allowing you to monitor and control the unit from anywhere.

Figure 3-1. My most recent reactor.

Figure 3-3 shows an awesome project I developed while living in Sebastopol, California. It's a reactor module that generates a healthy biofilm for hydroponic systems. I loved hearing from clientele how it boosted the vibrancy and health of their plants. It did this by making nutrients more bioavailable and breaking down waste residue, while simultaneously stimulating the plants' immune response.

Figure 3-2. This reactor (the same as in Figure 3-2) is powered by a Linux board running a Node.js application, which controls a custom sensor/actuator array and serves a local user interface.

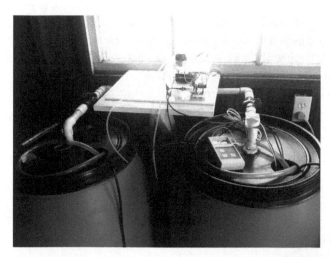

Figure 3-3. Reactor module that generates a healthy biofilm for hydroponic systems.

Figure 3-4 shows the doser controller for the hydro reactor. It optimizes the feed of "microbe food" based on dissolved oxygen and pH feedback. Before I started working with embedded Linux boards, I used an Arduino and a WiFly shield here to control the unit and send sensor data and commands to/from the Web.

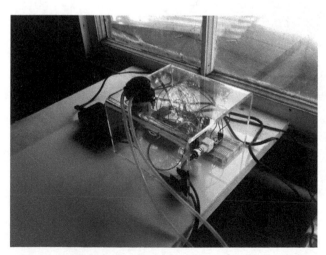

Figure 3-4. The doser controller for the hydro reactor.

The first reactor that I ever developed, seen in Figure 3-5, was in partnership with Jim McElvaney, a brilliant bioengineer, on Dan Smith's 30-acre organic farm in Sebastopol (see Figure 3-7). We brought in organic waste feedstock from around the community and digested it into organic, nitrogen-rich fertilizer for the farm. Inside the digester is a special biofilm matrix Jim developed to accelerate the digestion rate. We also used some interesting aeration techniques to culture facultative microbes (which can survive in aerobic and anaerobic environments). This limited volatile organic compound (VOC) production, leaving us with a healthier fertilizer.

Figure 3-5. My first reactor.

Figure 3-6 shows the digester control center, which was powered by an industrial programmable logic controller (PLC) along with a series of sensors and actuators (pumps, grinders, and a condensing boiler system). It was vital to controlling the temperature, pH, and redox of the system to ensure optimal conditions for the growth and stability of the digestion.

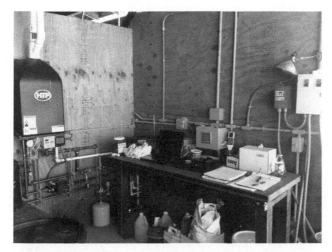

Figure 3-6. The digester control center.

Figure 3-7. The farm where I created my first reactor in Sebastopol, California.

Reactors can have far-reaching impacts on peoples' everyday lives and health. They'll lessen the burden of modern-day agriculture on the Earth, help us cycle our waste and nutrients, and enable us to do things that we can't even imagine yet. They're biological machines in their essence, yet when you really think about it, every living organism is a bioreactor, and in many ways our planet Earth is a reactor as well. Seeing a reactor through a more interconnected and biologically organic and malleable lens will help to harmonize the connection between biology and technology. It'll connect the best farmers in the world with the best engineers, and that's when we'll start making magic.

Gregory Mueller is a designer and embedded systems engineer working to bridge the gap between technology and the biological world. He has apprenticed at and managed farms in Italy, Argentina, and California, and is currently developing a project called Mu Enviro Reach (http://www.muenviro.com) (a control system that allows you to monitor conditions of a living environment). Reach him at gmono1@gmail.com or @gregorymu.

The Robotic Worm

Timothy Busbice

One of the age-old questions has been whether the way a brain is wired, negating other attributes such as intracellular systems biology, will give rise to how we think and how we behave. We are not at the point yet to answer that question regarding the human brain. However, by using the well-mapped connectome of the nematode *Caenorhabditis elegans* (*C. elegans*, shown in Figure 4-1), we were able to answer this question as a resounding yes, at least for simpler animals. Using a simple robot (a Lego Mindstorms EV3) and connecting sensors on the robot to stimulate specific simulated sensory neurons in an artificial connectome, and condensing worm muscle excitation to move a left and right motor on the robot, we observed worm-like behaviors in the robot based purely on environmental factors.

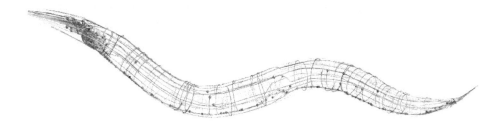

Figure 4-1. The C. elegans connectome, courtesy of NeuroConstruct (http://www.neurocon struct.org)

Our artificial connectome uses a program that can be started 302 times, where each program inherits the attributes of one of the worm's 302 neurons. These attributes consist of the neuron itself (named, for example, AVAL, DB02, and VD03) and the neurons that it connects to. We use the number of connections that a neuron has to another neuron as a weighted value. For example, if neuron A has three synaptic connections to neuron B, when neuron A "fires," we send a weighted value of 3 to neuron B. Using UDP[1] message communications, we can message the weighted values between each simulated neuron by assigning a port number and IP address.

Each simulated neuron program has a weight accumulator that sums the weights as they are received; a threshold is established that must be met before that neuron will fire. If no message activity is received within 200 ms, the accumulator is automatically set to zero (e.g., depolarizes the cell). Once a neuron fires, the accumulator is also set to zero. This gives our artificial connectome a temporal paradigm that has similarities to living connectomes.

To connect the robot to the artificial connectome, we created a program that reads the robot sensors every 100 ms. Depending on the sensor, we send weighted values to a specific set of simulated sensory neurons. For example, we simulate the worm "nose touch" by using a sonar sensor on the robot. If the robot comes within 20 cm of an object, the well-defined sensory neurons that are associated with nose touch on the worm are activated with UDP messaged weighted values. Likewise, there are many motor neurons within the *C. elegans* connectome that stimulate each of the 95 body muscles. Four rows of muscles are aligned down the worm's body: two rows ventral and dorsal left, and two rows ventral and dorsal right. We create a 4 x 24 matrix of these muscles, with each cell of the matrix representing one of the 95 muscles. We accumulate the weighted values on the left and right to drive the left and right motors on the robot. Motor neurons can be excitatory or inhibitory, and we send positive weighted values for excitatory synapses and negative numbers for inhibitory synapses. This, in turn, causes the two wheels on the robot to move independently, forward or backward.

---

1 The User Datagram Protocol (UDP) (*http://bit.ly/udp-wiki*) is one of the core members of the Internet protocol suite (the set of network protocols used for the Internet). With UDP, computer applications can send messages, in this case referred to as datagrams.

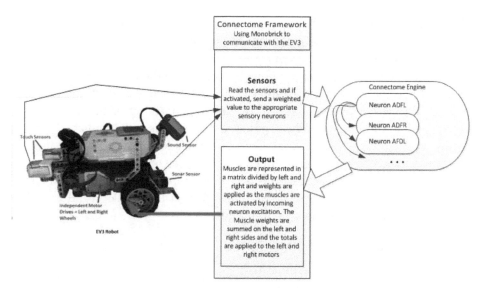

Figure 4-2. The Lego Mindstorms EV3 robot sensors are read by an Input program that activates the appropriate sensory neurons of the simulated connectome. An Output program receives the motor neuron output and accumulates weighted values to drive the robot wheels.

In general, the EV3 robot using the artificial connectome behaved in very similar ways to the behaviors observed in the biological *C. elegans*. In the simplest of terms, stimulation of food sensory neurons caused the robot to move forward. Stimulation of the robot's sonar, which in turn stimulated nose-touch neurons, caused the robot to stop forward motion, back up, and then proceed forward, usually in a slightly skewed path. Touching the anterior and posterior touch sensors caused the robot to either move forward (anterior touch) or move backward (posterior touch). *There is no programming to direct the robot to behave in any specific manner. Only the simulated connectome directs when the robot will move a motor forward, stop, or move backward.* This answers, at a very basic level, that the connectome alone gives rise to phenotypes that we observe in animals.

A YouTube video (*http://bit.ly/celegans*) shows the robot using the connectome framework and simulated *C. elegans* nervous system. The first part of the video displays the sensor input program that captures sensory data and sends it to a set of sensory neurons. This part of the video also shows the output program that captures the motor neuron weights; the weighted data is accumulated by the left and right sides of the *C. elegans* body muscle structure, and the accumulated weights are sent to the left and right motors of the robot. The middle of the video shows the robot as it comes up to a wall, activates the nose-touch sensory neurons, stops and changes direction, again totally under the control of the simulated

C. elegans nervous system. The last part of the video is a capture of the neurons as they are activated, showing green as weights are received and dark green when the accumulated received weighted value exceeds 10.

The *C. elegans* connectome is highly recursive. When the connectome reaches a sufficient level of stimulation, the connectome will continuously self-stimulate: a neuron (presynaptic) will stimulate another set of neurons (postsynaptic), and in turn, many of those postsynaptic neurons will stimulate the originating presynaptic neuron, creating loops of stimulation. The recursive nature of the connectome has shown to be a key factor in the connectomics research and resulting behaviors of *C. elegans*. This is now becoming a focal point of our analysis, to determine how these recursive loops play on the topology and resulting actions when the connectome is fully engaged.

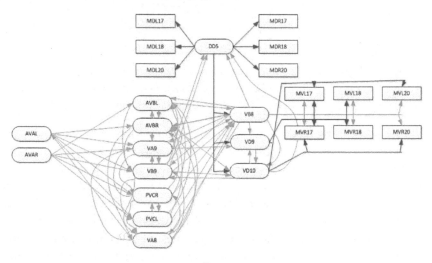

Figure 4-3. This image is created from the direct data output of the simulated C. elegans connectome, and specifically the network that surrounds the activation of neuron DD05. The green arrows are excitatory (positive) stimulation, and the red arrows are inhibitory (negative) stimulation. Oval shapes represent neurons, and rectangles represent muscle cells. As you can see, the network is highly recursive, whereby often neuron A excites neuron B, which in turn excites neuron A.

Although we can show that simple neuronal connections can give rise to expected behaviors, there is much more to the neurons of *C. elegans* (and of other animals) than just neuronal connections—including, but not limited to, the difference between chemical and electrical connections, neuropeptides and the various peptides and innexins that create neuronal complexities at the cellular level. Just the differences in chemical (synapse) and electrical (gap junctions) warrants the possibility of two programs to shadow one another and represent a single simulated neuron. Whether this evolves into multiple programs that together comprise a single neuron, or a single application that encompasses all of the systems biology of a single neuron, we must continue to improve and add complexity to get a true representation in reverse-engineering biology. This also includes the spatial aspects of how neurons are placed and connect throughout the nervous system to create a spatio-temporal model (which we have seen is important regarding body-touch sensing).

The artificial connectome has been extended to a single application written in Python and run on a Raspberry Pi computer. To our surprise, this simple program and version of the *C. elegans* connectome worked very well. We are currently creating a self-contained, Raspberry Pi–controlled robot that will be completely autonomous and independent of Internet connectivity. Our objective is to develop robots that can use the artificial connectome as a means to not only adapt to and navigate unknown environments, but also carry out specific tasks such as identifying or reporting environmental changes that could be vital to specific interests.

There is still very much to experiment with and analyze in reverse engineering the nervous systems of animals. We believe that this first step in being able to study an entire connectome, from sensory input to motor output, and the observations of expected behaviors will allow us to move forward in the understanding of nervous system wiring and how it develops into our behaviors. Moving from a simple 302-neuron connectome to higher-order animals will only increase the complexity and give us greater insight into how our own minds work.

Timothy Busbice is an independent researcher and founding member of the OpenWorm Project. Studying both computer science and neurobiology at the University of California, Riverside, has given Timothy a unique perspective in the balance between these two sciences. He firmly believes that reverse engineering the nervous systems of living animals will give tremendous insight into how the biology of connectomes work as well as give rise to intelligent machines.

Timothy can be reached at InterIntelligence Research, 869 Via Colinas, Westlake Village, CA 91362, by email at interintelligence@gmail.com, or on Twitter @interintel. Learn more about the C. elegans robot at www.connectomeengine.com.

A Glowing Trend

Glen Martin

Unlike many of his generational peers, Glowing Plant (*http://www.glowing plant.com/*) chief scientific officer Kyle Taylor (*http://bit.ly/mybio-kyle*) was never put off by genetically modified organism (GMO) crops. On the contrary: Kansas-born and bred, cutting-edge agriculture was as natural to him as the torrid summers and frigid winters of the southern plains.

"GMO corn first hit the market while I was still in high school," says Taylor, "And I have to admit I was fascinated by it. It was Roundup resistant, meaning that you could spray it with the most commonly used herbicide in commercial agriculture and it would remain unaffected. I found that really profound, a breakthrough."

Agribusiness felt the same way. Roundup (generically, glyphosate) is relatively benign as commercial herbicides go (compared to 2, 4-D and Paraquat, anyway): it binds to soil so migration to waterways is minimal, and it generally degrades quickly. Roundup-resistant corn—and later, resistant soybeans—ushered in the era of no-till agriculture. Farmers no longer had to cultivate between their rows for weed control; they could spray right alongside their standing crops without affecting them. This bolstered profits, and it also had environmental upsides: fewer passes through the fields with heavy farming equipment meant a big drop in fossil fuel consumption and significantly reduced atmospheric carbon emissions. Less cultivation also meant less erosion because the topsoil wasn't disturbed and exposed to rain and wind.

Taylor's interest in agriculture was still strong when he graduated from Iowa State University, where he earned his BS in agriculture biochemistry with an agronomy minor.

"But I got a bit disillusioned [with GMO technology] as an undergrad," Taylor recalls. "Initially, I viewed it as a means for feeding the world, for ending hunger and poverty. But I ultimately came to understand that famine is at least as much a political as a technological problem. I realized that this silver bullet mentality I'd been cultivating wasn't really grounded in reality."

Still, he didn't lose his passion for synthetic biology. The ability to manipulate genetic material struck him as—well, not magic. He was a scientist, after all. But it was wondrous, transcendent. So, he went to Stanford, where he earned his PhD in cellular and molecular biology.

"Maybe GMO crops weren't going to save the world, but molecular engineering was still one of the most powerful tools ever developed, and I wanted to contribute," Taylor explains. "I worked my butt off, and not because I expected a payoff. I was just thrilled to be involved."

After earning his doctorate, Taylor started playing around in DIY bio spaces. It was great fun, but he was looking for something more than amusement. He yearned for a mission, something that could properly accommodate his ambition and talents. Then he met Antony Evans (*http://bit.ly/ae-ted*), who had an MBA from INSEAD (*http://www.insead.edu/*) (motto: The Business School for the World) and worked as a management consultant and project manager at Oliver Wyman and Bain & Company. Evans had an irrepressible entrepreneurial streak. He had founded the world's first pure microfinance bank (in the Philippines) and developed a mobile app in conjunction with Harvard Medical School.

The two men got along swimmingly, and they realized their skill sets dovetailed. A partnership seemed logical. But what would they do? What could be their product or service? Taylor had been fooling around with bioluminescence in DIY bio labs, simply because he found the phenomenon charming.

"Bioluminescence is such a good teaching tool for biocoding in general," Taylor says. "When people take a black light and they see a plant glow, it moves them emotionally, especially if they had something to do with it. It opens them up to the possibilities of synthetic biology, its positive aspects, in a way that GMO crops can't."

Here, then, was their project: an open source company purveying bioluminescent plants and associated seeds and materials. It wasn't about ending hunger, but in its small way, it *was* about saving the world—or at least, it was about helping create a generation of bioengineers who might collectively solve some of the planet's thorniest problems. GMO corn and soybeans, Taylor observes, have been less-than-effective ambassadors for synthetic plant biology. To attract the young, the hip, and the brilliant, the discipline requires symbols and totems that speak to inclusion and harmony, not corporate dominance. And Day–Glo houseplants looked like a pretty good place to start: they seem self-contained, controllable, and friendly in a whimsical, Dr. Seuss kind of way. There is no intimation that they could escape from their pots and wreak environmental havoc.

"It started off as an educational platform," says Taylor of the duo's initiative, "And at this point, maybe it's more of a novelty. It's challenging. There's this strange tension we're dealing with, between teaching and inspiring on the one hand and making money on the other. There's certainly no getting around that second part, of course. We needed money to fund what we were doing."

To get those bucks, the partners went the route now in vogue with so many Silicon-centric start-ups. In April 2013, they launched the world's first DIY bio crowdfunding campaign on Kickstarter (*http://bit.ly/glowplants*), ultimately raising almost $485,000 from more than 8,400 investors.

Taylor and Evans consider Glowing Plant part of a continuum. Bioluminescence has long functioned as a basic biohacking tool, in that it illuminates, so to speak, cellular mechanisms in a relatively straightforward and easily comprehended way.

The first significant breakthrough in the field occurred in 1986, when researchers inserted a luciferase-producing firefly gene into tobacco plants. The plants glowed—albeit dimly, and only after extended exposure to light. In other words, they could not glow autonomously.

Fast forward 24 years. Researchers at SUNY announced the first "auto-luminescent" plant (*http://bit.ly/al-plants*). Rather than relying on firefly luciferase, SUNY's team inserted bacterial genes linked to luminescence into tobacco plant chloroplasts. Again, the light from the plant was muted at best, but it at least required no outside source to luminesce. At about the same time, a University of Cambridge team combined firefly luciferase and genes from the light-emitting bacterium *Vibrio fischeri* to create "Eglowli" (*http://bit.ly/eglowli*) bacteria, which emitted enough light to allow a close reading of the *London Times*.

All this research inspired a syncretic strategy for the partners. What could be achieved by combining the various approaches?

Quite a lot, as it turns out. Glowing Plant expects to ship its first products by the end of this year: potted *Arabidopsis* (rock cress) plants, ready for display next to your lava lamp, and packets of fertile *Arabidopsis* seeds. Glowing roses will be available in 2015.

"You can't use them as a desk lamp, but they definitely glow," says Taylor of the *Arabidopsis* products. "And we're getting more intense illumination with each generation. We're not at the 'tree night-light' stage yet, but we intend to keep turning out bigger and brighter plants. Will they sell in the marketplace? It's unclear at this point. But they're beautiful and fascinating plants. I'm cautiously optimistic."

Chemical Storage in DIYbio

Courtney Webster

When building your first lab, it's tempting to follow the same chemical storage system you had in grad school. But each school's unique categorization (e.g., Group A, Storage Group 02, and Code Blue) is like a foreign language to a local fire marshal or Environmental Health & Safety (EH&S) rep. Trust me—you don't want those inspections to be any harder than they have to be. Following a few simple guidelines will help keep your lab safe and inspection-ready.

Step 1: Assess

First, get a general sense of the chemicals you need (or will need) to store. You don't need to do a full inventory; just identify what requires refrigeration, how much flammable solvent (e.g., ethanol or isopropanol) you want to have around, and what can be stored at room temperature.

Step 2: Shop

If you have any flammable solvents, you should consider purchasing a *flammable-storage cabinet*. You can store a certain quantity (give or take 10 liters[1]) on the bench, but it's best practice to keep everything but your squirt bottles in a proper flammable-storage cabinet. You can get a flammable-storage cabinet on eBay (we certainly have), but make sure it has locking handles and, preferably, self-closing doors.

If you need any compressed gases (for example, CO_2 for a cell-culture incubator), purchase wall bracket(s) and chains for securing the gas tanks before you buy them. You'll want at least *two* chains for security: one around the middle-top and

---

1 The total volume you can store outside a flammable-storage cabinet depends on the type of flammable chemical and room (check out the *NFPA 30: Flammable and Combustible Liquids Code*) and any local guidelines.

one around the middle-bottom of each tank. These things are little potential rockets, so your local inspector will take their storage seriously.

 Your compressed gas provider can help you out by speccing brackets, chains, and regulators for you ahead of time.

If only media and aqueous buffer solutions need to be kept cold, any fridge or freezer will do. Just slap a big "No food or drinks" label on it. If you have more serious cold chemical storage, you can buy a combustion-proof fridge.

Finally, you'll want a bunch of simple plastic containers for secondary containment. This provides spill and earthquake protection and keeps your lab handy-dandy organized. Go on a shopping spree (treat yo'self!) to get the cabinets, containers, fridges, and shelves you need.

Step 3: Separate and Label

Now, divide and subdivide your chemicals as follows:

Group chemicals into liquids, solids, and gases.
Easy! So far, so good.

Subdivide by primary hazard into the following groups:[2]

- Flammable

- Corrosive

- Oxidizer

- Water-reactive

- Toxic

- Nonflammable, nonreactive

In a DIYbio situation (where you're keeping core go-to reagents but not necessarily a huge inventory), this gets you most of the way there!

---

2 "Chemical Safety in DIYbio," *BioCoder*, Summer 2014.

Subdivide again, based on chemical type.

For a small shop, you probably just need to separate your corrosive group into acids and bases. For more serious inventory, you'll need to be a bit more thorough:

- Organic bases (a base that contains carbon, such as triethylamine or pyridine)

- Inorganic bases (a base that doesn't contain carbon, such as sodium hydroxide)

- Organic acids (an acid that contains carbon, such as trifluoroacetic acid)

- Inorganic acids (an acid that doesn't contain carbon, such as hydrochloric acid)

- Oxidizers (for example, hydrogen peroxide)

- Pyrophoric and water reactives (in DIYbio, do you really have any of these?)

- Flammable solvents (ethanol, isopropanol)

- Nonflammable, nonreactive (sodium chloride, TEAA, SDS, Triton-X)

Voilà! Each group goes into a separate container (e.g., liquid organic bases, solid nonflammable/nonreactive, and so on).

This is where academic institutions try to simplify by giving each group a color, letter, or number. You should just label each container with the type to keep your local inspectors happy.

Step 4: Store your Containers in Cabinets and Shelves

Ideally, each group of chemicals would be stored in separate cabinets. This way, no chemical groups would mix together if an accident or natural disaster occurred.

If you don't have the space to store everything separately, you can store some chemical groups together if you keep incompatibilities in mind. This is where your alma mater system comes in handy, as a detailed description of what-can-and-cannot-go-with-what will be too long (and too boring!) for this article. I organized my lab based on the same system I used at Stanford University[3] (Figure 6-1), and that worked just fine.

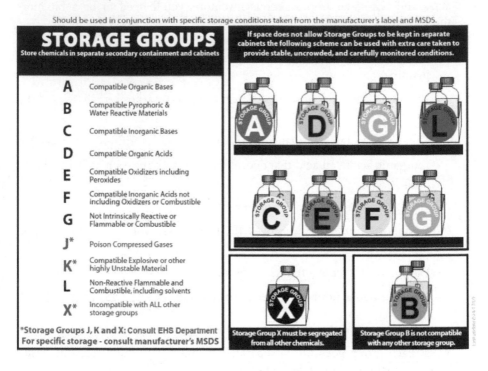

Figure 6-1. Stanford University chemical storage group system

Wherever you decide to put your chemical groups, make sure the primary hazard (flammable, corrosive) is clearly marked on the outside of the cabinet. This helps emergency personnel in the event of a fire or spill.

Inspections

Depending on your lab and the chemicals you store, you might require an inspection by the local fire marshal and EH&S. Stay tuned for a more detailed article on chemical permits, hazardous waste disposal, and what happens during an inspection. Until then, here are a few quick tips:

- Treat your first inspection as a learning experience.

---

3 See the Stanford University Compatible Storage Group Classification System (*http://bit.ly/chemstorage*).

- Make sure the lab is clean and tidy to give the inspector a good first impression. (What's that old expression? A cluttered room reflects a cluttered mind? You get the gist.)

- Be nice. Inspectors are used to people being defensive. Tell them what you're working on (in layman's terms) so they associate you with a story and have a sense of what you're trying to accomplish.

That's it for now! Stay safe out there.

Courtney Webster is a reformed chemist in the Washington, D.C. metro area. She spent a few years after grad school programming robots to do chemistry and is now managing web and mobile applications for clinical research trials. She likes to work at the interface of science and software and write for scientists and engineers. You can follow her on Twitter @automorphyc and find her blog at http://automorphyc.com.

SynBio and Environmental Release

Brian Berletic

When we consider the potential of synthetic biology, synthetic life-forms, and even entire synthetic ecosystems, we are first overwhelmed with the fantastical possibilities. Next we begin to work out the what-ifs, including, "what if our novel organisms make it out into the environment?" How might they interact, change, or adversely affect the environment?

While environmental release is one of many challenges facing those working in the emerging field of synthetic biology, the concept of invasive species working their way into our environment and wreaking havoc is a problem human beings have faced for quite some time. The snakehead fish, a voracious predatory species from Asia, has made its way into North American waterways. Its ability to survive even when the bodies of water it inhabits are dried out make it difficult, if not impossible, to eradicate. A combination of legislation and economic incentives to catch the fish are attempting to keep their numbers in check, but in all likelihood, the environment and the people who inhabit it will have to learn to coexist with this invasive species.

In other cases, biological management strategies have worked to contain and control invasive species. In Australia, the elm leaf beetle is an invasive species detrimental to Australia's elm tree (also imported). To control them, their natural predator, a parasitic fly known as *Erynniopsis antennata*, is used to prey on the beetles. Combined with best practices, less-toxic insecticides, and careful monitoring, the threat of the elm leaf beetle can be minimized.

The presence of invasive species can have a series of effects. The snakehead outcompetes native fish and could diminish their numbers or displace them entirely. Their biology, differing from those of native fish, invites the possibility of new parasites and diseases previously unknown and dangerous to native species. The elm leaf beetle was devastating elm tree populations before a system of management was devised.

Another threat comes from genetic pollution: invasive species, both natural and engineered, can begin combining with compatible native species, altering them genetically, reducing or eliminating desired, natural genetic traits. Biotech companies have faced stiff fines for accidental releases of engineered plants that persisted and genetically contaminated native species in the environment.

Laws already exist around the world, recognizing the threat of invasive species in attempts to prevent undesirable environmental releases. This can be seen at every airport or border crossing, where signs warn of bringing animals, produce, plants, and seeds into neighboring countries without declaring them for inspection. Extensive rules and regulations are also being established around the world, governing or entirely restricting the testing and use of genetically modified organisms to head off genetically engineered environmental releases that may become disruptive. The Cartagena Protocol on Biosafety is an international framework that seeks to safeguard natural biodiversity from potential harm posed by genetically modified organisms.

To assuage these growing concerns and to limit the threat of persisting genetically altered species in the natural environment, engineered organisms are being made that are either functionally sterile outside specifically controlled conditions; cannot survive beyond carefully controlled, artificial conditions; or both.

Dealing with Synthetic Invasions

As the tools of synthetic biology become more accessible to a larger number of people, the fear of environmental release will grow with it. However, the wide availability of tools that can sequence, analyze, manipulate, synthesize, and reintroduce genetic information into any given organism could increase our ability to cope with environmental release. Having more people increasingly literate in the skillful application of these tools could also help.

The designing phase of a synthetic organism might include steps taken to likewise ensure that it cannot exist or reproduce outside a predetermined, specific set of conditions. Like production-line fish used in aquaculture that are rendered sterile outside hatcheries, synthetic organisms engineered by individuals might include "contingencies" that prevent them from thriving in our natural environments. They may be engineered with deficiencies that require supplements they can receive only under controlled conditions. In the event of an accidental environmental release, without the ability to breed or address their built-in deficiency, they would quickly die off.

One example of this already in practice is that of genetically engineered AquAdvantage salmon. These fish are specifically engineered to be bred in tanks far inland and with specific physical characteristics designed to intentionally prevent environmental release. The distant proximity of their artificial habitats to natural salmon populations, the fact that they are engineered to be sterile as well as to survive only in water conditions that are carefully controlled, ensures that accidental environmental releases would be rare and unlikely to create havoc.

Another example, perhaps more relevant to synthetic biology, is GeneGuard (*http://bit.ly/geneguard*), described by its creators as "a modular plasmid system designed for biosafety." It works by encoding antitoxins on an engineered host bacteria, as well as toxins within the plasmids inserted into the bacteria, thus preventing lysis (cell death). In tandem, the host bacteria and the plasmids happily replicate, but in the event of horizontal gene transfer in the environment, natural bacteria without the antitoxin characteristic will die, and the toxin-laced plasmid transferred to them will not be replicated.

Of course, we must imagine that the more hands the tools of synthetic biology end up in, the greater the chance an accident will occur, releasing a novel organism into the wild with the ability to thrive, and perhaps even outcompete, native species. As is the case with invasive species today, management strategies could range from economic and legislative incentives, to the use of biological management. Barring the existence of a natural predator that could be used to prey on an invasive synthetic organism, a novel predator could be designed to not only target it, but to do so more precisely and efficiently than can be done with existing biological management strategies.

Already today, genetically modified individuals of invasive or harmful species are being produced to then be released into the wild as part of a biological management strategy. Mosquitoes developed by Oxitec in the United Kingdom are engineered to carry a lethal gene in effectively sterile males that is passed on upon mating with females in the wild. The gene, when expressed, subsequently kills the resulting offspring. Since the offspring receiving the gene die off before mating again, the management strategy is considered *self-limiting*, or in other words, sustained releases of Oxitec's mosquitoes would be required to continue the management strategy. Oxitec's mosquitoes are not designed to pass on their genes and establish themselves within the natural population, allowing the natural population to recover if the management strategy is abandoned.

Oxitec's solution for combating mosquitoes is a biological strategy for managing pests designed not to become an invasive pest itself. A similar strategy is being developed to combat invasive carp in Australia.

Handling Invasion on the Molecular Level

The proliferation of biotech tools used for synthetic biology may offer another set of challenges on the molecular and genetic level. While the battle against genetic pollution today is combated more or less the same way an invasive species is, with the tools of synthetic biology in hand, it may be possible to detect, monitor, and even reverse genetic pollution, if not altogether protect against it. The ability to sequence, analyze, rewrite, synthesize, and reintroduce genetic information is the basic process behind gene therapy and gene editing. Understanding what is wrong or undesirable about existing genetic sequences and introducing desired changes is what has allowed rare genetic conditions to now be corrected. Dealing with genetic pollution may benefit from a similar process.

The ability to translate DNA into digital code and back again through DNA synthesis opens the door to the possibility of genetic "hard drives" we can keep "offline" and secure in the case of an outbreak or environmental release. The ability to reintroduce that safely stored sequence into a corrupted genome through a process like gene therapy or gene editing could be developed into a genetic "reboot," providing a defense against poorly designed, unpredictable organisms or those engineered to be malevolent to begin with.

While people fear the worst regarding synthetic biology and the prospect of an environmental release, the double-edged nature of this technology may allow us to manage such releases better than we can with invasive species today. By ensuring that people both understand synthetic biology and have access to the tools necessary to control it, we will be able to preserve our natural genetic heritage as well as explore the possibilities, benefits, and potential pitfalls of future synthetic organisms and ecosystems—with relative safety.

Brian Berletic is a designer, writer, and intern at Desktop Genetics. He also runs a small makerspace/DIYbio lab called Helios Labs in Bangkok, Thailand. Follow him on Twitter @HeliosLabs or check out the Helios Lab blog (http://helioslabs.blogspot.com).

The Future of Food

Ryan Bethencourt

"So I had an accident."

That was the call I got from a scientist entrepreneur friend of mine, John, the CEO of Gene and Cell Technologies (*http://www.geneandcell.com*). He'd been working on potential regenerative medicine therapies and tinkering with bioreactors to grow human cell lines. He left the lab for the weekend, and then something went wrong with one of his bioreactors: something got stuck in it.

"So I was wondering what happened with my bioreactor and how this big chunk of plastic had gotten in there and ruined my cytokine production run. I was pulling it out, and I thought it was was weird because it was floppy. I threw it in the garbage. A little later, after thinking about it, I realized it wasn't plastic and pulled it out of the garbage."

What John had inadvertently done was grown a thick chunk of tissue, about the size of a small ear and about half as thick as your pinky finger.

He called me because he wanted to run an idea by me: Do you think we could use this accident to make burgers, if I could reproduce it? Would people eat it?

Yes, *in vitro* (lab-grown) meat has been the Holy Grail for many decades for tissue engineers, vegans, environmentalists, and many others who have looked upon the archaic and inhumane meat industry with a keen feeling that one day technology will revolutionize this industry, and it's coming. Groups like New Harvest (*http://www.new-harvest.org*) are supporting the technologies that will free animals from their use as food and still give consumers freedom.

What John did may be revolutionary, as some great accidents are, or may be hard to reproduce after all. So far, that accident has an N = 1, but if he's successful in replicating it, which he's working on, we may soon have lab-grown hamburgers, not in the $300,000 range but in the $10 range.

The exciting thing is, this is only the start of our post-animal future. It isn't just the archaic and inhumane meat industry that's about to fall to the power of science, but also egg farms and the cheese and dairy industries.

Hampton Creek foods (*http://www.hamptoncreek.com*) is creating plant-based alternatives to eggs, and they're already selling their super popular Just Mayo in Whole Foods, Costco, and many other places, with more products coming. Counter Culture Labs and Biocurious have successfully raised over 250 percent of target on a successful crowdfunding campaign to make animal-free cheeses (*http://bit.ly/veg-cheese*) from GMO yeast. Muufri (*http://bit.ly/muufri-milk*), a small, SOSventures seed-funded biotech company is developing animal-free milk.

For those of you who wonder whether you'd eat a lab-grown hamburger or drink lab-made milk, ponder this thought, which my friend Florian Radke shared:

> Imagine a clean lab production facility with pure sugar water and vitamins feeding the meat cell stock, with windows everywhere and accessible to the public, without the mess, blood, and guts of a standard slaughterhouse. Where would you prefer to get your food from?

Ryan Bethencourt can be found on Twitter at @ryanbethencourt. He heads up Life Sciences at the XPRIZE Foundation, is CEO and cofounder of Berkeley Biolabs, and advises multiple early-stage biotech companies. He's also cofounder of Counter Culture Labs, Lifespan.io, LA LAB Launch, SRG, and is a Life Sciences mentor at the Thiel Foundation.

How to Culture Biotech Startups in 100 Days

Sarah Choukah

Smells of yeast and *E. coli* incubating aside, something very pungent was in the air at University College Cork's microbiology lab this past summer. Synbio Axlr8r, the world's first completed synthetic biology accelerator, provided lab space, resources, and funding to six teams. The deal was simple: bootstrap a biotech company, build a working prototype, and do everything in between to have the startup take off in 100 days. The participants had all flown in from different countries—Austria, Canada, the United States, and France—with neat ideas and open minds. What they came up with in the end were not only amazing products, but a whole array of intriguing smells, uncommon flavors, changing colors, and novel textures and materials.

Ten or even five years ago, these projects would have been very difficult to develop for commercial applications in such short time frames. I'd like to offer some thoughts on what made this state of affairs possible in the case of Synbio Axlr8r: not a set of necessary conditions in particular, but a careful alignment of the right ones for the right teams. As more accelerator programs like Synbio Axlr8r pop up, expectations of what biotech and synbio startups can do are also going to change. The road from idea to final product will be much shorter, something that'll help the synbio consumer market establish itself, just like IT and software companies created consumer markets in the '80s and '90s.

SDE—Synbio Development Environments

An accelerator program can distinguish itself from others on two main fronts: the funding structure it offers and the environment the teams will grow in. Bill Liao, European business partner at SOSventures—the VC firm that founded and funded the Synbio Axlr8r—explains it in terms that emphasize context as the main determinant to the program's success. Creating a good context means dealing with a lot of uncertainty: whether the lab space, the people, and resources available to the teams are the right fit the first time around, for example. You can't add things up together and pinpoint a particular one that makes everything work when you describe context. Bill's strategy rather lies in carefully arranging initial conditions garnered through successive observations of SOSventures' other accelerator programs,[1] mostly software and hardware based until recently.

Those conditions added together were intended to make the accelerator as seamless an experience as possible. At the same time, it felt like a fast-paced, dynamic environment where scientists and researchers could develop several skills at the same time. Jacob Shiach, the program's director, didn't structure it in a stepwise or linear manner; it wasn't constructed as a series of blocs, each distinct from the other. Mentors came in every week to give presentations and workshops on several aspects of running successful biotech startups. The science and engineering themselves were crucial, of course. But we also got to understand how financials, marketing, branding, and team-building can make a great synbio idea into a successful, relevant, and transformative innovation. Weekly meetings with the program directors also gave us a better idea of how investment works within the startup ecosystem, what to expect from our first financing round, and how to prepare for engaging discussions with investors. We also experimented hands-on in creating and maintaining the presence—online and off—of our companies and teams and learned to value the importance of that presence. We used social media platforms as storefronts; creating content for them was a good way to reflect on the company's vision and the way we are presenting its goals as well as our accomplishments. Twitter (*https://twitter.com/HyasynthBio*), Facebook (*https://www.facebook.com/hyasynthbio*), LinkedIn (*https://www.linkedin.com/company/hyasynth-biologicals*), and AngelList (*https://angel.co/hyasynth-bio*) accounts as well as our website (*http://hyasynthbio.com/*) also helped in networking, interfacing with potential investors, and keeping in touch with mentors.

That presence is also necessary when it comes to getting closer to the people who are at the other end of the development pipeline. Startup structures are flexible in this regard; they afford the possibility to reach out to adopters and users early on in the product development process. Knowing about their needs right away instead of spending years developing a product improves its chances to survive after it's on the market. Teams get better at figuring out how to fine-tune projects and learn quickly from mistakes. That every member of my team is trained and used to rapidly prototyping genes and organisms also greatly helps in

1 Some of these include their most popular: Haxlr8r (*http://www.haxlr8r.com/*), Selr8r (*http://www.selr8r.com/*), and its most recent one, Food-X (*http://www.food-x.net/*).

this regard. This also comes naturally to us as the best way to go about conducting business: being fast and learning to gage success by achieving smaller milestones, one at a time, instead of big and lofty goals.

Culture and Cultures

A successful startup project is not just about making awesome things, but also about finding the right words and concepts to describe and present them. This matters insofar as those same concepts form the mental and cultural environments people live in. In the same way, new technologies are always developed or "packaged," so to speak, with ideas that make their novel aspects more familiar and easier to understand.[2] This is also how, within the program, each team got the opportunity to develop its own biotech startup culture and vision.

For teams like Muufri (*http://www.muufri.com*), that work on biosynthesizing vegan milk with yeast, it's important to reach out to people with concerns over how milk is made. This involves pointing to a problem the team perceived very accurately: how do we produce milk that is both healthy and sustainable, considering that intensively farmed cows live miserably and produce methane, which contributes considerably to global greenhouse gas emissions? Briefcase Biotec (*http://kilobaser.com/*), a company of three Austrian microbiology and computer engineers from Graz, works on a DNA synthesizer that emulates the ease of use you'd find in a desktop printer. Their DNA synthesizer is to biotech what the personal computer was to software and hardware development. University College Cork's iGEM team, Benthic Labs (*http://benthiclabs.wordpress.com/*), also participating in the program, works on expressing protein from the hagfish—an ugly living fossil that produces thick slime with unique properties when it's threatened by predators. In proposing to make biodegradable polymers from those proteins, they envision turning the plastics industry into sustainable manufacturing models.

At Hyasynth Bio we culture cannabinoids, compounds usually extracted from cannabis in *Saccharomyces cerevisiae*. Cannabinoids have high value in therapeutics and can help treat an impressive array of conditions: multiple sclerosis, chronic pain, and Alzheimer's disease are only three of more than thirty currently covered by Health Canada's medical marijuana prescription program. Being from Montreal, we couldn't fail to notice changes in the regulatory landscape—in Canada and elsewhere—when it comes to the ways cannabis gets defined and used in medical practice. The plant, although used for millennia in different kinds of ther-

---

2 An example of this is skeuomorphs and among them, app icons that link together old forms of a technology with a new one. The phone app icon is a phone receiver most of the time. The computer's graphical user interface is a desktop, and the address book icon most commonly looks like the book form of an address book.

apeutic contexts, is difficult to standardize for medical use today. It can contain various proportions of different active compounds that also make it hard to develop prescriptions for specific diseases. In having yeast optimized to produce controlled quantities and pure yields of compounds, we're using a powerful combination of engineering techniques honed through decades of development in software, hardware, and the life sciences.

Figure 9-1. From left to right: Sarah Choukah, cofounder and CEO of Hyasynth Bio, Synbio Axlr8r's intern Laura Eivers, and Hyasynth Bio cofounder Alex Campbell.

With this combination, we aim to develop new pharmaceutical standards based on novel biosynthesis techniques. Cannabinoids are only precursors for a wider array of applications for us: as more genomic data is available on various indigenous plants and organisms that have yet to be discovered and sequenced, we're looking for plant analogues that could do the job of producing even better compounds than cannabis, or improve cannabinoid formulations through hybridization with other kinds of plants.

We call cannabinoids *cultured* because we ferment and culture them with sugar, water, and organic precursors, for one. Second, we inscribe our technology within a continuous cultural landscape, one that transcends cultures yet knows different iterations in every one of them. What we explore and iterate on as a biotech startup, like several other Synbio Axlr8r teams, are ways to displace our common understanding of notions such as *organic* and *natural*. Culturing microorganisms has always implied culturing ourselves as humans at the same time. It also calls on a whole register of terms that are much more familiar to us: in a sense, we ferment therapeutics, just like Afineur (*http://hello.afineur.com/*), a team composed of Sophie Deterre and Camille Delbecque, ferments their coffee to radically improve its sustainable production. Revolution Bioenginering (*http://www.revolutionbio.co*), headed by Keira Havens and Nikolai Braun, is all about creating appealing flowers that change color throughout the day. Their flowers stand as examples of how synbio and aesthetics can challenge the way we think about the most familiar items in our everyday lives.

Figure 9-2. Kevin Chen, COO and cofounder of Hyasynth Bio

And Down the Road...

Founders of a biotech startup need to be confident in their ideas and vision. But they also need to be sensitive to doubt and know how to turn it into a particular kind of criticism. I'm not talking about the kind of critique one would find in a movie review or a judgment based on personal opinion, but something altogether different: a critique that would help entrepreneurs, scientists, and researchers make their actions more relevant in regards to long-term goals and improve the rationale of their projects.

Good mentoring is essential to cultivate that kind of attitude, regardless of whether it is offered within the context of an accelerator program or without it altogether. I was lucky to find myself surrounded by peers and mentors who also value the importance of good critique and showed high expectations of everyone else in this regard. The more than 100 days we spent together building biotech startups felt far from the glamorous, overhyped, and naive ways entrepreneurial lifestyles can be described. Culturing life in a petri dish, especially within the context of a biotech startup accelerator program, can't be dissociated from culturing the life skills and attitudes that make the whole setting possible in the first place.

Sarah Choukah is currently completing her PhD in communication at the Université de Montréal in Québec, Canada. She is also cofounder and CEO of Hyasynth Bio as well as cofounder of Bricobio, Montreal's first community biology laboratory. Sarah is also a Genspace alum member and NYC Resistor alum hacker-in-residence.

Community Announcements

Bay Area Science Festival

Come check out the Bay Area Science Festival from October 23 to November 1, 2014! Events include Explorer Days, where you can learn about the science of bread and cheese or hawk migration and branding; hands-on science at the farmer's market; or Discovery Days in AT&T park or North Bay, where there are hundreds of science activities for folks of all ages. Or maybe you want to attend *You're the Expert*, a public radio program using comedy to increase the accessibility of academic research. For an up-to-date calendar of events at the BASF, please see the website (*http://www.bayareascience.org/calendar/*).

SynBioBeta Conference 2014

Register for the third annual SynBioBeta conference being held in San Francisco from November 13th to 15th. The purpose of this conference is to gather researchers, investors, policy makers, and thought leaders to discuss the latest commercial advances in synthetic biology. Consider attending if you are part of a startup, in industry, an academic, an investor, a tech scout, or a biohacker in the field. Check out the schedule and speaker list (*http://bit.ly/synbiobeta*).

Indie Bio

INDIE BIO

After a successful first class, in which six teams started with just an idea and launched funded synthetic biology startups, Indie Bio (*http://indieb.io/*)is expanding in scope to include all bio-related startups and will be running classes year round, starting in January 2015. Classes will be taking place in the new Indie Bio Lab space that is currently being built in downtown San Francisco. If you have an idea, from synbio to biomed to bioinformatics, apply today to receive $35k in seed funding.

London Science Festival

LDN SCIENCE
FESTIVAL 2014

Check out the London Science Festival from November 12^the to 19th. This festival brings biosciences to the public, encouraging people of all ages to engage in scientific debates, learn more about current research, and consider pursuing careers in various related fields of study. Many of the events are free and cover topics such as fish oils synthesized in genetically modified crops, animal locomotion, painless injections, biofilms, scaffolds and cells, and much more. For an up-to-date list of events please go to the official website (*http://londonsciencefesti val.com/*).

Lightning Source UK Ltd.
Milton Keynes UK
UKHW021836040821
388201UK00005B/53